Not a Dog

Claudia Guadalupe
Martínez

Illustrated by
Laura González

Charlesbridge

In the desert grasslands, where a sign in the shape of a **flecha** points to Llano de la Soledad, a sprawling colony lies underground.

A perrito llanero is born there.
This so-called prairie dog is Not a Dog.

It is a tiny hairless thing, nuzzling its mamá with the **triángulo** of its nose. It needs her milk to grow.

When the perrito is two weeks old, its fur comes in. At four weeks old, its eyes open. At six weeks old, it walks.

Its appetite also grows. The milk and fresh grass that its mamá brings are not enough.

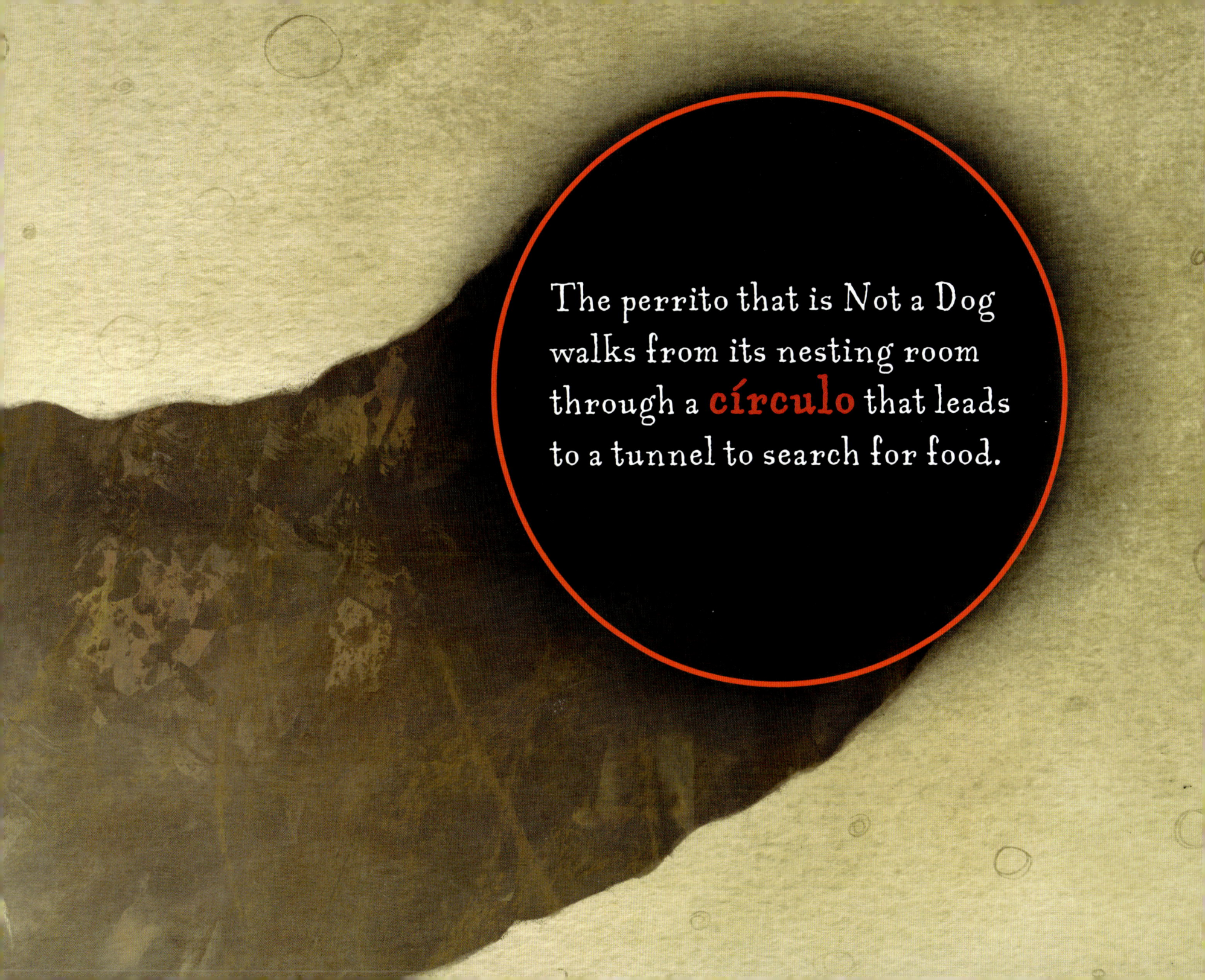

The perrito that is Not a Dog walks from its nesting room through a **círculo** that leads to a tunnel to search for food.

On the other side, there is a **cruz** where four tunnels connect.

To the left is a room for sleeping, to the right is a bathroom, and straight ahead is a room full of other pups. The perrito walks toward them.

Yip! Even though it barks a greeting,
it is still Not a Dog. It is a rodent.

One of these pups hugs the perrito.
Their bodies form a **corazón**.

Another pup digs into the soil.
Like all other prairie dogs, the
perrito digs to extend its burrow.
This also aerates the soil.

The pups eat grass and other plants with their sharp **rectángulos** for teeth. When they poop, the poop becomes fertilizer.

Meanwhile the pups watch for the menacing *figuras* of the coyote and *tejón* hunting side by side.

A large shape darkens the ground. It is the shadow of a hungry águila in the sky. The eagle spreads its majestic wings and swoops down.

Yip! The perrito barks a warning.

The pups jump over a rock that looks like
a **rombo** and escape into their burrow.
They are safe!

Óvalo-shaped eyes peer through school bus windows. The amigos and their maestra are on the lookout, too.

According to the teacher on
the school bus, prairie dogs once
built colonies from here to Canada.

These colonies fed and sheltered many animals, including chorlos, tecolotes, and zorros.

Then the farmers and the ranchers came into the prairie with their **luna**-shaped equipment. They destroyed the colonies, thinking the perritos were a threat to their crops and pastures.

Without the perritos, the **estrella**-flowered hediondilla bush pushed its roots into the soil, sucking up the water.

The prairie withered,
and the desert closed in.

Protecting the perritos
will protect many other animals
and save the shrinking grasslands.

The amigos wait, watch, and learn as the horizon cuts the sun into a **semicírculo**.

Back at school, the amigos pull out their papel **cuadrado**. They write letters.

They spread the word to the ejidos
and plead on behalf of the perritos.

Now, where a sign in the shape of a flecha points to Llano de la Soledad, the perrito that is Not a Dog is not alone.

As far as those óvalo-shaped eyes can see, there are prairie dogs and prairie dogs and prairie dogs.

VOCABULARIO: VOCABULARY

ÁGUILA: eagle

AMIGOS: friends

CHORLOS: plovers

COYOTE: coyote

EJIDOS: community farms

HEDIONDILLA: creosote

MAESTRA: teacher

MAMÁ: mom

PAPEL: paper

PERRITO LLANERO: prairie dog

TECOLOTES: owls

TEJÓN: badger

ZORROS: foxes

FORMAS: SHAPES

CÍRCULO: circle

CORAZÓN: heart

CRUZ: cross

CUADRADO: square

FLECHA: arrow

ESTRELLA: star

LUNA: crescent/moon

ÓVALO: oval

RECTÁNGULO: rectangle

ROMBO: rhombus

SEMICÍRCULO: semicircle

TRIÁNGULO: triangle

AUTHOR'S NOTE

The Mexican prairie dog (*Cynomys mexicanus*) is Not a Dog. It is a rodent closely related to a squirrel. Its name comes from the bark-like sounds it makes. Scientists believe these barks form an extremely complex language. A bark can sound the alarm or contain a description of predators. Barks can even describe shapes!

The natural lifespan of a Mexican prairie dog is three to five years. Female Mexican prairie dogs give birth once a year to four or five pups. All species of prairie dogs (black-tailed, white-tailed, Gunnison's, Utah, and Mexican) are born hairless and with their eyes closed. They can walk around and find their own food by six weeks. The plants they eat are also their primary source of water.

Prairie dogs live in networks of burrows, called colonies. These colonies once spanned the North American continent—the largest known colony measured 25,000 square miles!—but the prairie dog population declined greatly as people moved into the area, especially during the early- to mid-twentieth century.

People are still the biggest threat to prairie dogs. Ranchers worry that prairie dogs compete with their cattle for pasture, and farmers fear the prairie dogs will eat their crops. Prairie dogs have been targeted as pests and killed. Because of this, Mexican prairie dogs are now endangered.

In response, the Mexican government designated Llano de la Soledad as a protected area in 2002. Llano de la Soledad—which translates to "plain of solitude"—has the largest longstanding population of prairie dogs in Mexico. It is home to more than fifty prairie dog colonies. Prairie dogs are not alone here. Their community includes a number of animals and plants that depend on them—making the prairie dogs a keystone species. They are food for predators, share their burrows with other small animals like burrowing owls, and maintain the prairie's vegetation.

In order to further protect prairie dogs, local conservation groups have established sustainable grazing practices for cattle. However, potato farming still looms as a threat. One way to help is by supporting snack-food makers and grocery stores that source potatoes sustainably. Educating others about the perritos can also help make sure that prairie dogs remain protected.

To L. O. again, for fighting the good fight—C. G. M.
To my family—L. G.

At publication, all URLs in this book were accurate. Charlesbridge, the author, and the illustrator are not responsible for the content of any website.

Charlesbridge • 9 Galen Street, Watertown, MA 02472 • www.charlesbridge.com

Library of Congress Cataloging-in-Publication Data
Names: Martínez, Claudia Guadalupe, 1978– author. | González, Laura, 1984– illustrator.
Title: Not a dog / Claudia Guadalupe Martínez; illustrated by Laura González.
Description: Watertown, MA: Charlesbridge, 2025. | Audience: Ages 3–7 | Audience: Grades K–1 | Text in English, with some Spanish vocabulary. | Summary: "Learn about the endangered Mexican prairie dog. Includes back matter with Spanish vocabulary and an author's note about conservation efforts."—Provided by publisher.
Identifiers: LCCN 2024012481 (print) | LCCN 2024012482 (ebook) | ISBN 9781623543044 (hardcover) | ISBN 9781632899460 (ebook)
Subjects: LCSH: Mexican prairie dog—Juvenile literature. | Mexican prairie dog—Conservation—Juvenile literature.
Classification: LCC QL737.R68 M37 2025 (print) | LCC QL737.R68 (ebook) | DDC 599.36/7—dc23/eng/20240731
LC record available at https://lccn.loc.gov/2024012481
LC ebook record available at https://lccn.loc.gov/2024012482

Printed in China • OPIC
(hc) 10 9 8 7 6 5 4 3 2 1

Illustrations created in traditional media and Photoshop
Hand-lettering of title by Laura González
Text type set in Aunt Mildred by MVB Design
Designed by Diane M. Earley
Production supervised by Jennifer Most Delaney